# WOODSPIRITS

# WOODSPIRITS

WRITTEN BY ELLEN FORT GRISSETT

FOR TOM CLARK WITH ILLUSTRATIONS

BY PAMELA RATTRAY BROWN

PUBLISHED BY CAIRN STUDIO, LTD.
DAVIDSON, NORTH CAROLINA

# CONTENTS

# A NOTE TO THE READER

The first, and largest, part of our book chronicles the friendship that began a number of years ago between Tom Clark, an artist and college professor, and Hyke, a woodspirit. And while many of Tom's questions concerning woodspirits were answered by Hyke in those early days—even as we have set them down here, in words and drawings—other facts about woodspirits have only recently been revealed.

In fact, Tom has learned so much about woodspirits since he first sculpted Hyke, seven years ago, that he has been eager to share that information with those who love woodspirits, too.

For those of you having questions about woodspirits after you read the tale of Tom and Hyke, we have included, at the back, a brief guide to woodspirits. In this guide you will discover where the word "woodspirit" came from, what a "cassone" is and why there is one in every woodspirit house, and other useful tidbits of information.

We expect these tidbits to only whet your appetite for more, and we have warned Tom that further accounts of his friendship with woodspirits may be needed.

Until then, however, we hope this book will serve as your door to the wonderful world of woodspirits, a world that Tom has captured so successfully in clay and that we have attempted to capture in word and illustration.

Ellen Fort Grissett, Writer
Pamela Rattray Brown, Artist

# WOODSPIRITS

On a chilly, damp, gray day near the end of winter, more than five but less than ten years ago, Tom Clark was grieving for his dog, Shady.

He stood at the head of Shady's newly dug grave, his hands jammed into his coat pockets. Despite the warm sweater he was wearing under his coat, a cold wind still continued to creep down his back. But Tom barely even noticed: he was too busy thinking about his beloved dog. He and Shady had had a long, loving relationship: indeed, Shady, when she died, had been fifteen years old, which in man's system of counting would have made her 105 years old. Tom had owned Shady—or rather, in the way pets have, she had owned him!— for most of her lifetime. They had shared good years, long years, companionable years.

During the weeks when the nearby college was in session, Shady used to follow Tom up the winding gravel road from his house to the two-laned paved road that led into town. Here she would wag her tail and stand watching him as he bicycled (if the weather was nice) or drove (if the weather was bad or he was feeling somewhat lazy) into town to teach his theology classes. Tom knew, because neighbors told him so, that Shady would watch him until he was out of sight, then amble back down the driveway to the house.

In the evenings, Tom would find Shady waiting patiently for him at the head of the drive, and she would spring up happily when she saw him and run ahead of him to the house, where they would have supper and pursue the usual evening activities: grading papers and reading for Tom, gnawing a bone for Shady.

And now she was lying in this grave, wrapped in Tom's favorite Harris Tweed jacket. He had picked the jacket for her shroud because it was his favorite—comfortable, cherished like Shady. It seemed the only thing he could give her, finally, that would do her sufficient honor. And now, though he knew deep down that tears would help, he couldn't seem to find any. He was empty, bereft, alone. Never before had his chosen life of bachelorhood seemed so bleak.

"Remember the good times you had with Shady," a little voice within him said. "She gave you all her love, and she knew you loved her. Few human relationships are so free of misunderstanding, stresses and strains as the relationship you had with your dog, and for that you should be glad. Grieve for her awhile, then go about your business, thankful for the years you had with her."

Tom shook his head, wanting the little voice to be quiet. "I'm miserable!" he shouted. "I don't have anyone now!"

"Then you must begin to look about you,

for there are plenty of creatures God has made who would comfort you and provide companionship for you," said the little voice, stubbornly. "You might begin with me, you know."

Tom realized with a start that the little voice he had been hearing had come not from inside but from something else—some one else—who was with him at Shady's grave. He looked about, wildly, and saw nothing but leafless tree branches, Shady's grave and a cold, gray sky above.

"I'm imagining this," he said to himself. "Grief has made me hear things that aren't really there." And he felt even sorrier for himself than before, if that was possible. He stayed by the grave, shoulders hunched and eyes painfully dry, and his mind was filled with images of Shady, especially Shady as a puppy.

It had been a beautiful day in April when a former student and his fiancée drove up the circular gravel drive in front of Tom's house. Tom had been cutting the grass, and so he didn't hear the car approach. They called to him and he stopped his work, delighted to see them.

The student had said, smiling at his fiancée, "I've been telling her that your yard is too beautiful and spacious for you to have it all to yourself, so we've brought you a companion, Tom."

"A mail-order bride?" Tom asked, laughing. Then he caught sight of a tiny ball of black-and-white fur bobbing in the grass at his feet.

"Well, it's a lady, all right," the student answered.

Tom scooped her into his arms, smiling as the puppy licked his face. "She's a shady lady," he told his friends, "judging her parentage by her looks!"

The puppy, now christened "Shady," licked Tom's face again. And the couple had laughed and said, "We now pronounce you man and best friend."

Soon Tom couldn't imagine life without Shady. She even went to the college with him sometimes, usually at the beginning and end of each semester and right before major holidays, when he was feeling festive and there wasn't much routine for a dog to disrupt. She had been a big hit with a lot of people in town, especially Mrs. Nichols, who had her capable hands full overseeing the college dormitories. She always said it was refreshing to spend time with a creature who wasn't stubborn, untidy or lazy. Shady was a hit, too, with the children of other professors. They used to come running across the quadrangle when Tom and Shady appeared, and Shady would be so happy her tail would almost wag her small body.

Tom sniffed. There was a lump in his throat now.

"Go ahead and cry if you want to." It was the little voice again, and again Tom looked around.

"Who are you?" he finally managed to ask aloud.

The voice replied, sounding faintly amused but sympathetic, too. "Don't be embarrassed. A lot of humans don't see us at first. We're

very small, and sometimes we're hard to spot."

"Who is 'we'?"

"We woodspirits," the voice answered. "I'm going to move a little, and if you feel comfortable getting down on your knees and looking at me, I'll let you see me. But take your time, take your time."

Tom, feeling like a fool, carefully dropped first to one knee, then the other. He felt the knees of his trousers getting a little damp, but he looked around anyway. There was the grave, there were the azaleas, and the small dogwood, and . . .

He saw, finally, a slight movement among the damp, brown leaves at one edge of Shady's grave. First a tiny foot in a leather boot was extended from the leaves, then another. Then came a pair of (short) legs encased in sturdy-looking cloth, then a round belly in a green coat with lapels made of leaves and two small (but nevertheless muscular-looking) arms.

"Can you see me now?" the voice asked Tom. "Or perhaps I should say, 'Do you want to see me now?'"

Tom couldn't find his voice, so he nodded, figuring that a magical little creature like the one in front of him could decipher a nod perfectly well. Nothing would surprise him anymore.

The leaves moved again and finally there was a round, wrinkled face under a funny little cap made of a leaf. Tom found himself looking straight at a most unusual little figure.

The figure laughed. "It shocks humans less to see me in stages rather than all at once, don't you agree?"

Tom nodded. He still couldn't speak.

"My name is Hyke," said the creature, "and I, too, loved Shady." He came out of the pile of leaves in which he had been taking a nap (or so Tom assumed, since the creature looked a little sleepy) and stood looking up at Tom, who was still kneeling in wet leaves. "I used to hide behind trees and jump out and shout at Shady when she was walking around, and she loved it. That dog almost wore me out playing that game!"

Hyke stood about a foot tall. His eyes were bright and dark . . . and his face! Tom thought he had never seen a face so full of good humor and character (and wrinkles) as Hyke's.

"How did you know Shady?" Tom asked. It seemed the most logical thing to say under the circumstances.

The little figure looked over at the newly dug grave and smiled slightly. "I have lived in these woods for many, many years, and so I saw Shady when she was first given to you, as a puppy. My, what a bushy tail she had!" He laughed, and the tiny sound seemed to swell until it filled the clearing where he and Tom stood.

"I never saw you."

"Well, we don't usually show ourselves to humans unless there is some way in which we can help them," said the creature. He smiled up at Tom again and settled himself on a log near the grave.

"And, too, you were usually in a bit of a hurry," Hyke added.

Tom nodded. "I suppose so," he said, thinking back to evenings when he had been so tired he'd done nothing but sit in front of the fire and scratch Shady's ears.

"We woodspirits admire the energy of humans," said Hyke, "but when you know us better you will see that being in a hurry is not usually our way."

"You are a woodspirit?"

"Certainly. What else could I be?" Hyke asked with a twinkle in his eye that even Tom, so much larger, could see.

"Well," said Tom, slowly, "you could be many, many things. I've read about gnomes, and fairies, and elves. . . . I thought you were one of them."

Hyke laughed. "Certainly not! Once you know us better, as I said before, you will realize that we are very different from elves and fairies. And from gnomes, too. . . . I know there was a very popular book written not long ago about gnomes, but woodspirits and gnomes are different, though we are distantly related, I suppose you could say."

pamela rattray brown

Then he broke off and looked up at Tom. "But this is not a good time to talk about elves and fairies and gnomes. You are grieving for a noble beast, as I am, and my purpose in allowing you to see me is not to frighten you but to help you. I know you feel very alone right now, and it does not seem as if time will ever be the healer that some say it is."

Tom nodded, unable to speak, and fished in his coat pocket for a handkerchief. He felt as if the lump in his throat was about to move up into his eyes and make him cry, something he rarely ever did. He started to explain how he felt to Hyke, but when he looked down he realized that the little woodspirit was gone. Tom looked around again, called Hyke's name a few times, then gave up. The woodspirit had vanished.

As the days passed, Tom's grief began to ease and he found himself full of curiosity concerning the little creature he had met—he thought—in the woods. Where did Hyke live? How was a woodspirit different from an elf or a gnome? And why had he chosen to appear to Tom? Now, whenever Tom walked along the lanes around his home, he was alert, watching for Hyke to pop up out of the leaves or appear on top of a rock. And when Tom left class every afternoon, he headed eagerly back to his studio, where his fingers dug into the clay he kept on hand and began to form a shape that resembled that of the woodspirit.

But Tom finally reached the point where he realized that his statue would be no good until he saw Hyke again. He had the figure right, and the colors of the garments, and he was confident that his artist's memory had accurately sized up Hyke's funny little hat. The hat looked pretty good, in fact. But the face!

Here Tom confessed to utter failure. No matter how hard he tried to remember Hyke's face, he couldn't seem to capture it in the clay. He began to pace up and down in his studio, wondering what to do.

"If I could only see him again!" Tom thought to himself. "Just one hour with him, and I could finish this statue!" But he also admitted to himself that what he wanted even more than the chance to finish the statue was another conversation.

Besides, he was just plain curious about woodspirits.

One evening, just as dusk was falling, Tom heard the faint ringing of his studio doorbell. He looked up from his work and called, "Come in, it's not locked!"

"You know perfectly well that I'm too small to open this door by myself!" answered a little voice, half with amusement and half with exasperation.

Tom jumped to his feet, ran to the studio door and flung it open, a big smile on his face. Hyke regarded him with twinkling eyes from a very precarious perch: he had somehow pulled an old stump from the yard over to the door and had been able to climb up on it until he could reach the doorbell.

"It's you!" Tom exclaimed.

Hyke smiled. "That is correct. You are looking and feeling much better, I see. I could certainly use a cup of that beverage you humans call tea, you know."

"Come in, come in, of course," said Tom, somewhat flustered.

"If you would be so kind as to give me a lift to the ground, I can come in," Hyke reminded him. Tom was delighted to do so—why, the little fellow was more solid than he looked, though still no longer than the measurement of Tom's arm from elbow to wrist—and he led Hyke through the studio to the back, where he always kept a pot of water ready to heat and some Earl Grey teabags. Tom busied himself with preparing the water while Hyke explored the studio.

"Are all humans this untidy?" Hyke asked with a smile.

"Well, many of them are, I suppose," Tom replied, a little embarrassed. "You see, I am an artist and a teacher, and the two together tend to make me a pretty bad housekeeper, I'm afraid."

Hyke nodded. "I, too, am not the best housekeeper," he confessed, "and I am a bachelor, besides, so I do not clean up very often for anybody. A guest, occasionally, or when a neighbor has a party and cannot provide space for all his guests, and so my spare room is volunteered!" The woodspirit chuckled. "And young woodspirits who want to get away from the fetters of school and family often come in and raid my cupboards . . . they know that I'm a very indulgent old stick, not likely to tell their parents on them."

Tom nodded. It sounded very much like his own bachelor life. In fact, he had already allowed one college student to live in a spare room downstairs and another one (graduated, but still unemployed) to live over the studio. Then there were the students who bicycled out to his house to ask him questions about term papers, and friends from town who brought him cakes and cookies they'd baked.

"I suppose you already knew I was a bachelor . . ."

"Shady told me," Hyke said with a nod, and Tom—totally captivated by his little visitor—didn't think that the idea of a dog and a woodspirit discussing his living habits was odd at all!

Tom poured a cup of coffee for himself, then looked about, wondering what he could find to hold Hyke's tea. After all, a human-size cup was hardly the thing for a woodspirit!

"A shell would do quite nicely," Hyke told him, pointing to a jar that held seashells Tom had gathered on his last trip to the beach. Tom fished about in the jar, finally coming up with a scalloped shell that could hold more than a mere mouthful. He filled the shell with tea, then handed it to Hyke, who grinned.

"Why don't you go back to work, and if you'll give me a lift again, I'll just sit and watch," he suggested. Tom placed Hyke at one corner of his artist's table and then settled himself on his stool, gazing at Hyke in wonder.

"Am I dribbling my tea or something?" Hyke inquired.

Embarrassed, Tom shook his head. "No, no, I'm just so amazed . . . so glad . . . so curious . . . " He stopped, then pointed to his half-finished clay statue. "I've been trying to make a statue of you, and I couldn't get your face quite right, and I was hoping you would come visit, and now here you are!"

"Woodspirits come when they are most needed." Hyke, holding the edges of the shell in his hands, sipped the hot tea carefully.

Tom thought about this for a moment. He had the faint stirrings of an intuition, something that told him that Hyke knew he, Tom, needed to be needed. . . . "How do you know when you're needed?"

"We just know, that's all," said Hyke, adding, "I can't stay here all night, so if you want to get my face 'quite right,' as you said, you'd better get started!"

Several minutes passed as Tom worked. He had always loved working with clay, but this

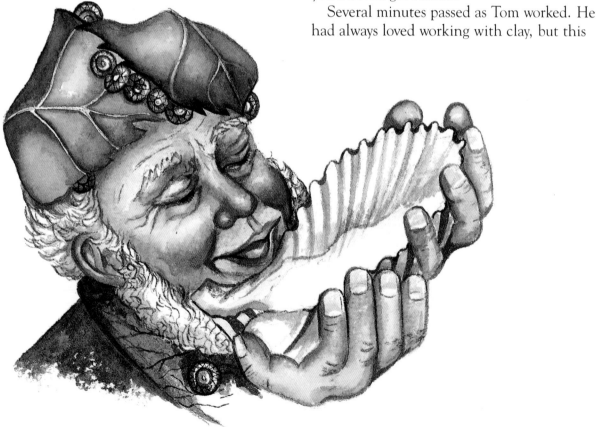

time he almost felt as if his fingers were shaping and molding and pressing smaller bits of clay onto the larger mass by themselves. They seemed to know what they were doing, and so Tom relaxed and let his fingers move. And after a while, the bits of clay on top of Hyke's clay figure began to resemble Hyke's face, wrinkles and all.

"Your face is wonderful!" Tom told the woodspirit. "Do all of you look like, well, you?"

"Some have more wrinkles, some have less," Hyke said with a smile. "We live a very long time, you know. Whereas one year of a human's life equals seven years of a dog's, one year of a woodspirit's life equals four or five years of an average human life."

"Amazing," said Tom, working away. "Humans hardly ever reach 100—more likely, just 70 or 75—yet you folks get to be . . . "

"Four hundred years old, in your terms." Hyke put his shell down very carefully and looked around him. "I see that you do what are called 'portrait busts' of people, do you not?"

"I've done quite a few," Tom answered. "But to be very honest with you, I was ready to try something new, and so that's why I began to sculpt you. I wanted to make something to go at the head of Shady's grave, and somehow a bust of a human just didn't seem quite the thing."

"Well, I'm honored. What else do you do? I mean, when you leave in the morning carrying all those books?"

"I teach religious art at the college," answered Tom. "And I serve on some committees, and I go to college functions and sometimes to the store. Fairly often I drive into the city for a meal, or a concert, or an appointment of some kind. And for recreation, let's see, I garden, and go for walks, and spend hours at the library. . . . Sometimes, on weekends, I drive several hours to see my family."

Hyke smiled at Tom. "I told you the first time we spoke that someday you'd see the differences between humans and woodspirits. I suppose that doing as much as you can as fast as you can is the human way. . . . Woodspirits have tasks to do, as you have, and we set aside special times for thinking or walking or reading, but our pace is much, much slower."

Tom nodded. "Sometimes I must look pretty silly rushing around. But I think humans often feel we have so little time on earth to do the things we need or want to do that we tend to— well—get caught up in the rush instead of enjoying ourselves. We don't live as long as you do, after all."

Hyke laughed. "That is true. But humans might live longer by going at a more relaxed pace, more like woodspirits do."

Tom frowned. "But you don't get very much done when you move too slowly . . . "

" . . . That depends upon what you are doing and why," Hyke told him. "We con-

sider the time spent thinking about nature just as important as the time spent, say, cleaning our houses or baking bread. No woodspirit looks down on another woodspirit because he is sitting on a rock thinking. Take me, for example. I do my tasks each day, but I make time first and foremost for listening to music, walking among the flowers, reading the words of great philosophers . . . ”

“So woodspirits have philosophers?” asked Tom, intrigued. Somehow he hadn't pictured the merry, mellow Hyke as being one who enjoyed reading the profound and sometimes obscure thoughts of philosophers.

“Oh, my, yes. We have been around a long time, after all, and that means we have set aside eons of time in which to formulate theories and think deep thoughts.” Hyke said this with a grin, leading Tom to wonder just how serious his little friend was. “We have many philosophers going back centuries past your Plato or Socrates. In fact, in school we had to memorize the names of the ‘Great Twelve Hundred,’ as they were called.”

“Don't tell me,” said Tom, laughing in turn. “You had eons of time in which to memorize names, right?”

“Of course,” Hyke answered, smiling. “Woodspirits have very good memories, as you can imagine. Since we live so long, we have more to remember and more time to perfect the skill of remembering. In fact, the first years of woodspirit school are spent savoring the past deeds and thoughts of other woodspirits, storing those memories where they are readily accessible to us.”

“So you go to school.”

Hyke looked amused. “How many years do humans attend school? Twelve, isn't it? Plus, for those who wish it, four more years of college?”

“That's correct,” said Tom. “Then some people go even further. I myself went to school for, let's see, five more years following my four years at college.”

“Well, woodspirits go to school for most of the first one hundred years of their existence.

“Just think about it. It makes sense, does it not, when you realize that woodspirits live so very much longer than most living creatures?” Hyke asked. “Think about me, for example. I've seen many, many things that humans have only read about in their history books.”

Tom was amazed. “That's right, I hadn't thought about that. If you're 400 years old . . . ”

“No, no, I'm not 400 yet!” Hyke laughed. “I am in middle age, like you. About 300.”

“ . . . then you were around before America was settled, or the Revolutionary and Civil wars were fought, or the light bulb was invented, or the car . . . ” Tom stopped working and stared at Hyke, who had resumed his seat on the table. “How amazing!”

Hyke simply smiled. “Woodspirits are amazing. But then, so—in different ways—are humans, because they are the ones who have actually set all this history in motion. Look at your own father's lifetime. He went from gas lights to electricity, from horse-drawn vehicles to automobiles, from a primitive airplane to a man landing on the moon. And all of this has taken place in a hundred human years or less! We woodspirits are simply amazed."

"Hmmm," said Tom. "Then we should be proud of all that we've gotten done, shouldn't we?"

"Yes, you should," said Hyke. "But most humans I've seen have been so busy accomplishing things that they haven't stopped to marvel at the things man has done over the centuries. . . . As you yourself said before, Tom, the process of doing things sometimes becomes more important to humans than *what* they're doing."

Tom had to agree. He realized that lately he'd been so busy making daily lists of things to do that he hadn't stopped to wonder if those things were even worth doing! He had become very, very good at being organized and checking off tasks as he accomplished them, but were they worth accomplishing?

"You're right, Hyke," he said with a sigh, after thinking about all his lists. "It's been hard lately for me to make time for people or activities that I haven't written on my daily schedule . . . and sometimes those people have really needed for me to break my routine and spend time with them."

Hyke nodded. "Friends shouldn't be given only the time you happen to have left over at the end of a day, but often that's what seems to happen."

"Well, what do woodspirits do to make time for their friends?"

Hyke jumped up from his seat and replied, "When we see another woodspirit, we immediately name a day when we want to see each other again, and we both remember, and that day comes and we are so happy because we will be with our friend again."

He looked at the statue, which was coming along quite well, and nodded. "Very like me, I'm sure. But now I must go. Woodspirit families always take supper together when the sun has finally gone down behind the hills, and I can't be late. It's considered very rude to be late for supper."

Tom was dismayed. "But I'm not finished yet! And I haven't asked you half the questions I wanted to ask you! Can we do this again soon?" And then he laughed, hearing himself as Hyke heard him. "I mean, let's set a day and time when we do this again."

"Fine," said Hyke cheerfully as he clambered down from Tom's desk and dropped lightly to the floor. Woodspirits were quite agile, Tom noticed, despite their roundness. "Would tomorrow at this same time suit you?"

"Perfectly," said Tom, smiling down at Hyke. "This time I'll leave the door open for you."

[21]

And thus was born the relationship between Tom and Hyke. At first, they met every other evening in the studio, where Tom would attempt to capture Hyke's expression in clay. After the statue was finished, however, he was reluctant to say so, for fear that the visits with his little companion would end, too. But Hyke (sensitive to human feelings as woodspirits are sensitive to all living things) assured him that the visits would continue, whether the statue was finished or not.

One day Tom went to place the statue of Hyke at the head of Shady's grave, and he realized to his astonishment that spring had come. There were yellow crocuses peeking through the grass, which had once been brown and frozen but now showed signs of becoming green again, and there was a freshness to the wind that spoke of blue skies and

sunshine ahead. Tom had been so engrossed in his statue and in talking with Hyke that the cold, lonely winter months he had dreaded had passed, and he was peaceful again.

He squatted in front of Shady's grave, holding his statue of Hyke and thinking about how Shady would have loved to see spring come again to the North Carolina mountains and hills. He found that he could think of Shady with love and longing, yet without that all-consuming rage and grief that he had first experienced when she died. He felt very close to Shady, and very sure that she would like having a statue of Hyke standing guard over her, season after season and year after year.

In fact, Tom realized with a start that he felt as close to Hyke as Shady must have felt. "Hyke was very, very clever," he told himself, "to pose for my statue of him. He told me he had come because I needed him, and I did. Yet somehow, I feel that maybe he needed

me, too. Maybe I'm important to him, just as he is to me." It was a cheering thought, as the woodspirit had no doubt known it would be, and so Tom went home almost happy, convinced that someday he would see Hyke again.

As a matter of fact, Hyke appeared that very night in the studio. Tom, busy over a new lump of clay, didn't even hear him enter. All of a sudden, he realized that his wood-spirit friend was climbing up the rungs of an extra stool to the spot where he could make a funny little leap and land on the worktable.

"I went by Shady's grave on the way. The statue looks quite nice, if I do say so myself," said Hyke. Tonight he was dressed in hues of blue and purple and looked quite dapper. He peered at the lump of clay in front of Tom.

"What are you doing now?"

"I'm going to make another statue of you," Tom answered. "I enjoyed myself so much making the first one that I want to start again."

"Well, how about making a statue of another woodspirit besides me?" asked Hyke, smiling at Tom's excitement. "Yes, I believe that soon you must meet some of my kinfolks, as you humans call them. But first, I want to tell you a woodspirit secret."

He looked very shy, Tom thought, which was unusual for such an outspoken and cheerful woodspirit. (Though Tom was later to learn that in matters of a personal nature all woodspirits are somewhat reticent, feeling that too much self-absorption is unhealthy.)

"You're looking quite the gentleman tonight," Tom added, then smiled. "Gentle woodspirit, I should say."

"Well, today is the day before my birthday," Hyke explained, nudging the lump of clay with his boot, "and our custom is to begin several days before all our special ceremonies to wear brighter clothes and a bigger smile. It helps us prepare ourselves for such happy circumstances. And that is my secret."

"Well, congratulations!" Tom exclaimed, putting out his hand. Hyke shook his thumb carefully and sat down, then jumped back up again as if he couldn't sit still.

"I hadn't planned to do this," Hyke began, "but . . . would you do me the honor of visiting my home tomorrow night?"

He added hastily, "Not for a party or anything. . . . When you get to be my age, you are content to remember the wilder par-ties of your youth, and simply celebrate another year by yourself, with none of that fanfare you used to have. But I would like to be with you, Tom."

Tom was touched. "Of course I will come! I know how you feel. At my age, friends begin to forget birthdays, and any way, it's harder to find gifts that are useful rather than insulting."

"That is true," Hyke agreed. "I should be off now, for I want to attempt to put my house in order for the occasion, but I will come back tomorrow night at this time and show you how to find my house."

He laughed as he clambered down the stool rungs. "Of course, you have often walked right by my house! But I will take no

chances that your sense of direction—which is already quite horrible—will desert you tomorrow night."

Tom chuckled as he watched the sturdy Hyke leave the studio, then became curious: Where in the world was Hyke's house? And how could he have passed it and never noticed it before?

Tom was waiting in the studio the next evening, dressed in his best cord trousers and several sweaters (for the nights were still a bit chilly sometimes, this being early April) and with a present for Hyke in his knapsack: a very small sample of a wine made by a winery in the foothills of North Carolina that he thought Hyke might like. Tom had had to rummage through the shelves of the local dimestore to find the doll's bottle in which to pour the wine, and had suffered the surprised looks of the teenaged girl at the cash register, but the effort, he decided, had been worthwhile. He knew that woodspirits sometimes made their own wine from plants and stems, and he wanted Hyke to have a bottle of human-made wine on his table, even though there might be woodspirits who, like many humans, didn't drink.

After much thought, Tom had also put his sketchpad and pencil in his knapsack. He rarely went anywhere without them, and besides, who knows? Perhaps he would be able to make some new sketches of Hyke for another statue.

When Hyke appeared, he was wearing the brightest outfit yet: a magenta jacket, orange trousers, black boots and leaf hat, with a violet pinned to his lapel. The violet was so large by woodspirit standards that it resembled a big grape fastened by magic to Hyke, but Hyke didn't seem to mind.

"Follow me!" Hyke told him. "But mind you walk carefully," he added, eyeing Tom's walking shoes. "Even though we woodspirits can move as quickly as chipmunks when there is danger about, I'd rather stroll tonight, thank you very much!"

Tom discovered that "strolling" in woodspirit terms was about the same as strolling in human terms. For the first time in a long while, he was able to relax and look about —though he took care to stay a few feet behind Hyke, lest he flatten the poor woodspirit on his birthday!

He recognized many landmarks: There was the old stone wall built almost a century ago by some other farmer; here was the creek that snaked through his property and wound on through the hills in search of a river; there was the spot where he found a Jack-in-the-Pulpit last spring. He was amazed at how adept Hyke was at climbing rocks, picking his way through snarled bushes and avoiding holes in the ground, though he did help Hyke over the stone wall (Hyke told him

with a grin, "Usually woodspirits take the long way around this wall, but I counted on you to pick me up this time!").

Occasionally as they went along, Tom heard Hyke call out greetings of some sort—he assumed to other woodspirits in the area, but he did not hear any reply. After a few shouted messages, Hyke in fact stopped and shook his head in bewilderment. "I don't understand why no one's around!" he muttered, almost to himself, before he resumed his walk. Tom, behind him, saw the little woodspirit shake his head several times more as the two of them journeyed.

The sun was just beginning to drop in the sky, casting a pink and gold glow over everything it touched, when Hyke, followed by Tom, pushed his way through the overgrown vines at the base of a tall stand of trees and they emerged into a cool, quiet, mossy clearing.

"This is Sink's Farm!" Tom exclaimed. "I found this several years ago! It's quite well known to the 'locals' around here . . . "

Hyke nodded and led Tom over to a tree whose twisted, gnarled roots had broken through the ground centuries ago. Vines had grown over the roots, too, with the result being a leafy, green wall about three feet high. Tom had noticed it once, but without much curiosity. Now he was amazed to find that behind the green wall lay Hyke's house.

"How wonderful!" he told Hyke, leaning way over so he could look into the house, which was made of more branches laced

together and tied with small pieces of twine. The house was open on one side, with the tree roots forming its three other walls, and so Tom was able to watch Hyke move around inside his house.

It seemed to Tom to be very much a bachelor's abode, not unlike his own, as a matter of fact. There were austere yet solid-looking chairs carved out of oak in a style that somewhat resembled the nineteenth-century Shaker style. Lamps were made out of blown glass globes (learned from those masters the Viennese, perhaps? Tom wouldn't have been surprised) surmounted with fluted paper shades, possibly cut down from French candy papers. There were braided rugs on the floor, as well as what appeared to be a painted floorcloth of the sort that was popular in Colonial days in America. Some walls Hyke had left in their natural state, so that they were a warm, earthy brown from the sticks; others he had evidently plastered with some sort of woodspirit concoction: they were smooth, and slightly pink in tint, and the perfect backdrop for Hyke's pictures. Some of these seemed to be old stamps framed in wood; others were tiny renditions of woodspirits and woodspirit pastimes. (One scene in particular could have been an inspiration for Bruegel's *Children at Play*, minus the more ribald games.)

"Hyke, did you paint these?" Tom asked.

Hyke reappeared and looked pleased. "No, my father did. . . . Unfortunately his talent wasn't passed on to me." But Tom was too busy mentally calculating the age of the paintings—good grief, they had to be at least 600 human years old!—to offer any sympathy.

Tom had another surprise when he spotted Hyke's bed. It was a wonderfully intricate piece, combining what looked to be cherry wood with scrolls and flutes and columns and posts. The headboard and footboard, in fact, were painted in wonderfully vivid colors like pink, and blue, and green, and deep red, and some equally intricate letters were painted on the side rails with gold leaf paint. Tom supposed the letters must spell out Hyke's name in the ancient language that Hyke had told him about once (woodspirits

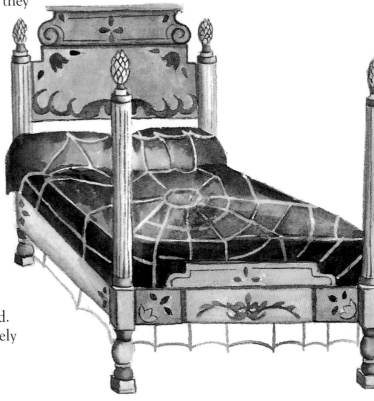

no longer spoke that language, but to be able to read and write it was considered a special sign of favor). The bed was covered with a spread of what looked to be—and in fact was—the most gossamer and lacy spider webs, preserved by the woodspirits with some special substance.

And everywhere Tom looked, he saw books! Books in woodspirit language, and Greek, and Latin, and French, no doubt copied from the human originals by wood-spirits of those times, and no doubt very valuable, if they were only big enough for humans to read! Tom had to smile: his own dwelling was practically knee-deep in books, papers and magazines, too.

Tom, in fact, felt right at home as he gazed down at Hyke's domain. Everything was comfortable, cozy and—thanks to Hyke's recent house-cleaning endeavors—clean.

"Your house is wonderful!" exclaimed Tom, enchanted with Hyke's ingenuity and amazed, too, at how like his own home the little woodspirit's dwelling was. He suddenly felt less out of place in Sink's Farm than he did when Hyke first led him there: somehow, when one saw a new acquaintance's home, a deeper tie was formed and a just-born friend-ship was strengthened. Or that is the way Tom felt about Hyke and his cozy little home.

Hyke seemed pleased by Tom's compli-ments on his house. "This is really an ordi-nary house by woodspirit standards," he said modestly. "All woodspirit houses are above the ground, unlike the tunnels gnomes prefer to live in, and some of them are quite

wonderful! I've seen them built high up in the trees, or carved into the side of a mountain, even ones floating upon the water like a houseboat!"

The woodspirit then disappeared into the kitchen, where he seemed to be busy with some pots and pans. In a minute, he poked his head out of the open side of his home and looked up at Tom, who stood nearby.

"You know, Tom, you might be more com-fortable sitting down than standing up all evening," he told his human friend, smiling. "I was just getting us something to drink, and then we will have a nice, cozy chat."

He disappeared again. Tom looked about the clearing and spied a soft, moss-covered rock big enough to accommodate his posterior; there was even a silvery smooth birch tree standing over the rock, close enough for him to lean against when he sat down. He settled himself comfortably, thought a moment and got out his sketch-book and pencil.

When Hyke appeared a few minutes later, Tom was busily sketching Sink's Farm and those parts of Hyke's house that he was able to see from where he sat. He was so engrossed, in fact, that he didn't hear Hyke until the woodspirit had coughed politely.

"Would you care for something to eat and drink?" Hyke asked him.

Tom realized that his trek through the fields had indeed made him both hungry and thirsty. "That would be wonderful," he told Hyke; then he stopped and shook his head.

"I am becoming forgetful these days! I brought you a present!" And he fished around in his knapsack, finally coming up with the doll's bottle of wine.

Hyke looked pleased. "What a treat! Occasionally someone will make some special woodspirit wine, but that is only every hundred years or so . . . and we've always been a little wary of drinking except on special occasions, anyway. But today is special, so I thank you, Tom. Now I'll go in and get some woodspirit beverages and edibles for you to try."

Tom, smiling, watched Hyke scramble over the tree roots above his house and vanish down some stairs made from twigs laced together. Really, it was amazing that a woodspirit of Hyke's well-fed appearance could move so quickly! Then he began to sketch again.

The shadows began to deepen. The sun became a butter-yellow circle near the horizon, then began to drop, acquiring pink and red hues as it descended. Tom sketched on and on, forgetting everything except his desire to capture Sink's Farm on paper.

Only the sound of a twig snapping made him look up from his drawing pad. There, on the far edge of the clearing, stood another woodspirit, somewhat shorter than Hyke but no less brightly clad. Behind him were four other figures who were whispering to each other and giggling occasionally.

Fascinated, Tom held his breath, afraid to speak or move lest he frighten the woodspirits away. The mistiness of twilight made it difficult for him to see their features, but he thought he detected a present in one woodspirit's arm and a dish or plate of some kind in the hands of another.

The leader of the woodspirits turned to them and admonished them to be quiet; then, picking his way carefully over fallen branches and rocks, he and the others made their way across the clearing and down the steps to Hyke's house, where Tom lost sight of them. But he could hear the shout "Birthday greetings!" and the *squeak!* that evidently meant Hyke was taken by surprise.

There followed some moments of general hilarity; then Hyke began to speak, so softly that Tom couldn't make out the words. There was a gasp of "Human? Here?" And then someone else exclaimed, "He'll step on us!"

Hyke evidently was reassuring his friends that no, big as Tom was, he wouldn't step on any woodspirits. "Besides, he's not very big by human standards, anyway," Hyke added, which annoyed Tom just a little (he was somewhat sensitive about his small stature) and didn't appear to have made the woodspirits feel any better, either.

"Hyke, if I'd known a human was going to be at this party . . ." someone began, indignantly.

" . . . Skipper, if I'd known you were going to tell me who I could and couldn't have at my own birthday party . . . " Hyke actually sounded menacing until he laughed. "Come

HAPPY BIRTHDAY HYKE

on, let's not fight, today of all days! You'll like Tom. He's Shady's human."

The mention of Shady evidently made everything all right again, and pretty soon Hyke appeared at the top of the steps, followed by five very befuddled-looking woodspirits. They were very careful to stand far away from Tom, which made him smile. He should probably be more in awe of them than they of him!

"Tom, I would like for you to meet four of my special friends, who have taken it upon themselves to make a lonely old woodspirit happy on his birthday," Hyke said, looking anything but lonely and old. In fact, he was positively beaming as he led his friends over to where Tom sat. Evidently woodspirits liked surprise parties as much as some humans did!

"I would get up, but then I couldn't see you very well," Tom told them. "My name is Tom Clark, and I live on what used to be a farm, near the college."

"We know," said the female woodspirit. "The place where there are two ponds, and all of those lovely ducks!"

Tom was amused. "I suppose you and the ducks are friends, just as Hyke was a big friend of my dog's?"

The woodspirits nodded vigorously, and one of them, Skipper, added, "We have known those ducks for years. Ducks are very curious by nature, you know, and so when we began to pass through your fields on our walks, the ducks just had to get to know us! We've had some good laughs to see you roaring down the road in your car!"

Tom laughed, too. By now he was used to hearing comments about his treasured Thunderbird; in fact, his nickname around campus was "T-Bird Tommy," which amused him very much.

"Enough of these jokes about Tom's transportation," Hyke told his friends, though he was laughing, too. "Say hello to Tom properly, like you've been taught."

The four woodspirits bowed in unison, very solemnly, and then one of them—this one a male, of about Hyke's age—piped up and said, "Is it true that you're an artist?"

"That's Skipper," Hyke interrupted, "and the others are Eva, and Shorty, and Franklin, and Freddie . . . you're forgetting your manners, young fellows! And young maiden, I should add."

"Yes, I'm an artist," Tom answered. "Actually, I'm really a sculptor. In fact I've made an image of Hyke in clay, and we've become friends. But please, Hyke, let's not all be so formal. This is your big day, and I don't want to distract any of you from your celebration. I'll just stay where I am and sketch, and the rest of you carry on with the party."

That idea seemed to suit everyone except Franklin who handed Hyke a present, muttered a gruff "Beg your pardon" to the group and walked away into the woods. Hyke smiled and explained to Tom that Franklin wasn't being unsociable, merely extremely bashful and that the woodspirit seldom, if ever, stayed at a party for more than five minutes.

Meanwhile Eva and Shorty had gone into the kitchen and come out again several times with bowls (made of acorns cut in half) holding what looked to be berry cookies. Skipper and Freddie between them managed to drag some barrels made of hollowed-out gourds up the stairs; when the cords holding the halves together were untied, these shells proved to be full of a dark, rich liquid. One of the woodspirits explained that the liquid was made from the first violets of spring, still hidden from human sight by the damp leaves of late winter but accessible to woodspirits, who were quite adept at spotting the purple blossoms.

Hyke thoughtfully put several of these vessels near Tom's elbow, assuming that one mouthful would not be quite enough by human standards. The cookies, when passed, proved to be delightful concoctions made of nuts, berries, and wheat flour, made by Hyke that very morning and left to bake in the sun.

After a few minutes spent drinking and eating, the woodspirits suddenly realized that it was growing quite dark in the clearing. Indeed, Tom could barely see the marks his pencil made on the sketchpad anymore, despite the fact that the rising moon gave promise of particularly bright light later on. His offer of some matches with which to start a fire was met by surprised silence.

"We know all about matches," Hyke finally told Tom. "In fact, we are surprised to know that humans still depend upon matches for fire, since woodspirits have invented much more sophisticated ways of producing a fire, and more foolproof, too! But please, do build a fire if you would like . . . only remember that we are fairly small, and the flames started by humans are usually too hot for us to bear for long. And, of course, we don't want to start a forest fire!" Hyke chuckled. "That would be terrible for a woodspirit's reputation!"

Tom looked about, then spotted a pile of rocks in another part of the clearing, part of the old stone wall that ran around Sink's Farm. "How about if I build a wall of rocks around the fire so that I can still have some light but you won't get too hot? And here there are no overhanging branches to accidentally catch fire," he said. The woodspirits nodded enthusiastically.

Accordingly, Tom gathered a number of large rocks and piled them in a circle, then found branches that weren't too damp and placed them in a heap in the center (Hyke contributed some dry pieces of kindling from his own woodbox). Then Tom very carefully fished in his knapsack until he found the book of matches he kept for just such emergencies. He lit his match (after four tries, which caused the two smaller woodspirits, Shorty and Freddie, to giggle) and carefully touched it to the branches, which after a moment's hesitation began to snap and crackle and turn into golden tongues of fire.

The woodspirits let out a collective "Ah!" and sat down nearby, basking in the warmth that seeped gently out from between the rocks in the protective wall.

Tom settled back down on his stone, positioning his sketchbook until some of the warm, golden firelight was thrown upon the pages, and began to sketch again. Pretty soon the woodspirits, their birthday toasts made, began to forget he was even there. Their talk turned to Hyke, the honoree.

"Are you sure you weren't expecting this party, Hyke?" said Shorty with a twinkle in his eye. He and Eva and Skipper were sitting on some smaller rocks in the clearing, not too close yet not too far from the fire, which was now crackling and dancing. Freddie, Tom noticed, seemed to be asleep in an abandoned bird's nest left lying on the mossy ground nearby.

"Well, I did *hope* you four might come by tonight," Hyke retorted. "But I'm amazed that I'm still around! A woodspirit could drop dead, frightened out of his wits by the likes of you appearing behind his back!"

"Oh, go on with you!" said Eva, smiling. Unlike the female gnomes Tom had seen depicted in one of his books, she was remarkably pretty and slim, not dumpy and plain, and he liked the way leaves were entwined in her hair. "You know better than to think we would forget your birthday, you old woodspirit! Why, we've been working for days on this surprise party."

"That's right," said Shorty. "In fact, we

were so busy we didn't even hear you and Tom go by! A chipmunk brought us the news and my, how we hurried to get here and surprise you before you started feeling sorry for yourself."

He turned to Tom and explained, "Just as you humans often have family members who are special to you and support you no matter what, we, too, have families. Only we call them something different."

"What?" asked Tom. Then, remembering his manners, he apologized. "I'm sorry, I shouldn't be so curious . . . "

"Woodspirits live in circles, which we call 'eisteddfods,'" explained Hyke. "It is a word that has been handed down to us from our Scandinavian ancestors, and it really means a group or special gathering. All of us who live within several miles of this spot are in the same eisteddfod, and we are sworn to love, protect, help and nurture each other."

"We are also responsible for all the humans within those miles," added Eva.

"Responsible for?" Tom was amazed.

"Of course," said Shorty, reaching for another cookie. "We all have several humans we watch over, and when they need us, we appear to them, or we make things happen to

them that normally would take much longer to occur. That is why Hyke knew when Shady died, and came to you . . . and why other events have taken place in history that seemed to be sheer luck or good fortune." Shorty chuckled. "Usually there was a woodspirit behind it all."

Eva took pity on Tom, who looked more puzzled than ever. "Think about it. Think about the discoveries that were made by explorers who thought they had been blown off course, or by scientists who worked and worked on an experiment and then poof! A solution to their problem just seemed to appear out of nowhere. But we only intervene, speed things up, when a great number of our humans would benefit from what we do."

Hyke nodded. "Eva's right . . . because as you probably are beginning to know, Tom, woodspirits have a different concept or idea of time from you, and we don't like to tamper with time unless it seems, well, natural to do so. Occasionally we do smaller things, such as loosen the stems of fruit so it will fall to the ground in time to be picked, or paint the centers of daisies rather than waiting on the sun to color them, or light the tiny lanterns inside the fireflies . . . "

Tom was fascinated. "Do all of you do all these things? Or do you divide up jobs like humans do?"

Skipper laughed this time. "We love that word 'job,'" he said. "It's a fairly modern word, by your human standards, and we hear it used all the time—usually not in a flattering way, either!—by the humans we watch over. What do you call a job?"

Tom considered this question for a few minutes. "I suppose I would say that my job each day at the college is to teach. But humans also have another word, which is 'calling.' My calling is when I look at whatever talents I have and see how they can be used for the greater good . . . for the divine purpose for us all," he said. Warming up to the subject, he plunged ahead while the woodspirits listened respectfully. "When I view sculpture as a means of a making a living, it becomes just a job. But when I try to create something for the enrichment and enjoyment of others, it makes my sculpture more of a calling."

Tom stopped, a little embarrassed that he had been so carried away by the subject, but the woodspirits made encouraging noises and nodded their heads. He knew that they must have already been aware of how he felt, since they were so wise when it came to discerning human thoughts, but nevertheless they made him feel he had said something very important.

Eva, in fact, said, "Perhaps humans and woodspirits both have callings, though that is not our word for them. . . . There are certain things we each are better at than other things, and though we enjoy every moment for its own sake, we would be dishonest if we told you some moments weren't twice as enjoyable as others!"

Hyke smiled at her. "What Eva is trying to say, Tom, is that she, too, has something she loves outside of her home and her offspring. . . . They are important, but she also has a special talent that is of great use to the entire community, the 'eisteddfod.'"

He pointed toward his house. "You saw my bed, did you not?"

Tom said enthusiastically, "It is beautiful! I have rarely seen such craftsmanship, even in human furniture . . . " He stopped and blushed as the woodspirits began to laugh. "All right, I'll stop comparing humans to woodspirits! Obviously humans would lose in most categories!"

"Well, Eva is the one who makes up the paints that are used on the bed," Hyke explained. "We have a custom that when you come of age—which is about 200 years old—each young woodspirit makes himself or herself a piece of furniture that will be in his house as long as he lives, then will be passed along to a young woodspirit of his family. It is a way of making sure that certain crafts, certain customs, do not pass away into the mists of time, as so many human customs have. . . . These things we make are precious, and it is the duty of each eisteddfod to make sure that they are taken care of."

Tom nodded, thinking regretfully of things his grandfather and father had made that had been thrown away, or sold, or given to someone who didn't appreciate them as he should.

"I chose to carve a bed for myself when I came of age," Hyke continued, "and when I was done, Eva came and painted it with her wonderful colors and it was unveiled at the coming-of-age ceremony for all to bless. If you look closely, you'll see where my knife slipped quite a few times, but Eva made it look better."

Tom was intrigued. "What kind of paint do you use? How do you make your colors? I've had trouble getting the exact colors I want from the art supply store . . . "

Eva smiled. "I make my paint from a very, very old recipe, one that my mother and her mother and her mother before her have passed along to me. . . . It is milk paint, you see."

Tom knew, of course, that in Colonial America, milk paint was used quite often on furniture, and that objects still bearing traces of original paint were almost priceless today. Eva nodded, as if she knew exactly what he was thinking.

"Yes, woodspirits taught the early settlers how to make milk paint from curds of freshly skimmed milk, lime from the creekbanks and water," she told Tom, "and in fact some of the northern woodspirits showed the Indians how to make paint from salmon eggs and red cedar bark."

"How do you produce the wonderful colors?" asked Tom, remembering the vivid hues on Hyke's bed.

"Well, years of experimenting have taught us that clays produce reds, yellows and grays.

Some berries are quite colorful when mashed and left out in the sun for a while, and ground brick dust makes a wonderful deep red, which we particularly like. And I am always trying new combinations. . . . It's particularly fun because milk paint is almost transparent, as you would say. Light shines through it and around it, which gives the coming-of-age pieces a, well, sort of radiant look."

Tom looked at Eva with new respect, noticing for the first time how delicate yet true the colors of her garments were: the blue of lapis, the pale yellow of new cream. She had made the dyes in which her clothes were colored, he now realized.

"So colors are important to you. What else do you do?" Tom asked the woodspirits.

Hyke pointed to Skipper. "There is our weather foreteller, right there. He is the one who can read the signs that nature leaves all around us."

"My grandmother could always tell when rain was coming because her arthritis flared up," Tom said with a smile. "Is that what you're talking about?"

"Well, that's a little primitive," sniffed Skipper. Evidently he was many steps beyond Granny's arthritic joints, and proud of it!

But he was either too proud or too shy to explain further, leaving Hyke to do it for him. "Skipper can tell if rain is coming from seeing how low to the ground the barn swallows fly," Hyke told Tom. "And, of course, there are crickets . . . "

"I've always heard you can tell something about the weather from crickets, but I never knew exactly how it worked," Tom confessed.

"Well, you count the number of times a cricket chirps in fourteen seconds," Skipper explained, "and then add the number forty to that to get the temperature. Works, too," he added gruffly, but his eyes were twinkling. "That's very simple, though. . . . Just like the rule of counting the number of fogs in August to figure the number of snows coming the next winter, or squeezing the corn husks. A thick one means a cold winter. . . . My parents and their parents before them didn't know anything about weather signs, so I learned from an old woodspirit who used to sail a boat on the pond near here."

"That's why we call him Skipper," Eva put in.

"I wouldn't think that a pond was very informative," Tom began, but he was stopped abruptly by Skipper's somewhat indignant words.

"Not informative! Did you know that a pond is a home for thousands and thousands of plants and animals? That a pond has a life span, just as other living things do? That you can read the history of a pond in its layers of mud at the bottom—even a pond as old as a woodspirit gets to be—just like you read the time lines in a tree trunk? That whirligig beetles have eyes with two hemispheres, one that can see in the air and one that can focus on things in the water? That—"

"Enough, friend Skipper, enough!" said

Hyke, laughing. "You don't want to bore Tom the first time you meet him!"

"I'd like to see this pond with you, if I might," Tom told Skipper, and was rewarded with a big smile and an acquiescent bow. "To think I used to catch tadpoles in that pond and never looked any further! And I may have to start listening to crickets, too," Tom added.

Hyke spoke up again. "There are many things Skipper could show you, Tom, including a barometer he made that is much more accurate than most human ones. . . . But he seems to like the signs of nature better than the woodspirit-made one, don't you, Skipper?"

Tom next looked at Shorty. "What does 'Shorty' mean? That he was small when he was younger?"

Shorty shook his head and laughed. "Actually, it's because my favorite meal is strawberry shortcake, which I don't get often enough to suit me! And as for a job or a—what did you call it, Tom?—a calling, well . . . "

"Shorty is our writer," Hyke explained. "He is responsible for writing down the history of our eisteddfod, day by day, and preserving it on special disks, which we bury in specially chosen places so that woodspirits to come can look upon our words and know who we were."

Shorty, seeing Tom's look of surprise, laughed. "Oh, it's not like the stone tablets that Moses showed the Israelites," he told Tom, adding, "Yes, of course we've read the Bible! And as many of the other things humans are partial to reading as we can stand, though none is so wonderful as the Scriptures."

Hyke decided to clarify things for Tom. "It's not primitive at all, Tom. You see, we are quite skilled in typesetting, in calligraphy, in other forms of writing that humans are just now beginning to experiment with using 'word processors'—is that the right term? The disks that Shorty uses are copper, and he has devised a way of engraving on them that will not lose its clarity over centuries of time. He's also come up with a special code for these disks that should help any woodspirit read them, whether or not our language as we know it today still exists or not. . . . It's almost like the special raised dots that are used in Braille, which he also helped a human invent."

Shorty looked proud but said nothing, evidently preferring that Tom hear of his deeds from the voluble Hyke.

Hyke continued. "Shorty is the best one among us in linguistics. In fact, back in your early nineteenth century, he was the one who helped Sequoyah come up with the Cherokee Indian alphabet."

"I certainly remember hearing about that in school!" Tom exclaimed. "After all, there are so many Indians in this part of the country . . . or were."

Shorty nodded. "When the first settlers from across the ocean came to this land, of course, we came with them. And we discovered that there were about, let's see, 35,000 Indians in this particular state when we woodspirits journeyed here. . . . We were able to set up a way of communicating with them and so were able to help ease some of the problems when the settlers came—though," and he shook his head, "not nearly enough, unfortunately."

"What about the alphabet, though?" Tom

asked him. "It was quite sophisticated, if I recall correctly."

"That's right. There were eighty-six symbols, some of them coming from Latin and some from the Greek, which of course we woodspirits are familiar with, having lived for many years in those lands," said Shorty. "It was a difficult language for white men to cope with, but it seemed to open many new doors for the Indians."

Tom nodded. He remembered that the Cherokees had even been able to use the language to translate the Christian Bible and to begin a newspaper, back in the 1800s before they were driven out of their lands and made to wander westward.

"Now Shorty keeps records of languages such as the one he and Sequoyah devised, and he records our special ceremonies and comings of age and our conversations with the humans we allow to see us," explained Hyke.

Tom wasn't so sure he wanted his side of this conversation recorded! But Hyke was gesturing towards Freddie, who still slept on in the abandoned bird's nest despite the talking that had been going on around him.

"What do you think of Freddie?"

"Well," Tom said, somewhat embarrassed, "he's smaller than the rest of you, so I assumed he was younger. . . . But maybe he's actually older and needs his sleep?"

Eva snorted. "If he were an old one, he would be up chasing the youngest woodspirits instead of sleeping, for it is our custom to have the youngest given into the care of the oldest for a time. The oldest woodspirits teach the young ones, and in turn the parents see to it that the old ones have a comfortable life and are truly appreciated for their teaching."

Tom thought about that. "I like the idea of the oldest woodspirits teaching the youngest. . . . Sometimes we humans think that elderly people have nothing left to share with young people, or don't have the patience to be around them."

"Actually, I think being around young woodspirits keeps them active, and the young ones learn respect and patience themselves," Eva told him. "The old ones actually look forward to the time when they can stop working at their tasks and begin teaching, instead of dreading it . . . and the youngest ones learn special customs that they normally might not learn."

"But Freddie, here, is not old. . . . In fact, as you guessed, he is the youngest of us all!" explained Hyke, smiling and nudging Freddie with his foot. Freddie merely rolled over in the bird's nest and went to sleep again.

[45]

Eva chose this time to stroll across the clearing and go down the steps into Hyke's house. Soon she appeared with two acorns of dark, sweet juice, which she offered to the others. Skipper and Shorty, evidently thirsty from all the talking they'd done, accepted; Tom, to be polite, declined. As a guest—and a strange human one, at that—he felt a show of restraint would be appreciated. He went back to sketching while the woodspirits settled down again, near the fire.

"We should tell Tom about Freddie," Skipper reminded Hyke.

"Freddie uses his sleep to particular advantage," Hyke told Tom. "You see, all woodspirits like to sleep, because it is a part of life to be enjoyed, but Freddie actually accomplishes things in his sleep . . . or his dreams, to be more precise."

Tom was puzzled. "You mean he writes stories in his dreams, or makes up situations?"

"Not exactly." Skipper chuckled. "Freddie here invents things."

Hyke carried on. "Freddie has been inventing things for, oh, centuries and centuries. . . . Did you know that the yo-yo, which Americans in particular have enjoyed playing with in your lifetime, actually originated in ancient Greece? And that there are paintings showing European rulers back in the 1700s playing with them?"

Tom was amazed. "I thought they were invented back in the 1920s," he said. "That's when I've read they were especially popular . . . "

"Well, Freddie had the idea back then to introduce them to humans here in the New World, and so he took the old Greek and French ideas and worked on them until the yo-yo was much better and smoother than it ever was, and he showed it to a human who loved it and began to make them. And that's why so many children today still play with yo-yos. I daresay the children in your town have them . . . "

Tom nodded. "In fact, I think I have the one I played with back in my childhood in some box in my basement. What else has Freddie invented?"

"Well, he's still young by woodspirit standards—hasn't even come of age yet—but one of his best ideas came about, let's see, a hundred years ago. There was a man who lived at the top of a mountain in Tennessee and the man had a little hotel he ran. He started having a hard time getting people to come way up there to stay, and business was really suffering. So Freddie went to sleep one day and dreamed that a golf ball went rolling around that mountain on a special track instead of in the air as it had always been used before. . . . And so when he went to this man whose little mountain hotel was about to be taken away from him, the two of them invented miniature golf!" Hyke explained. "The man became quite, quite wealthy from that golf course on top of the mountain, but he never forgot to be kind to woodspirits."

Tom looked at Freddie with respect. "I wonder what he's inventing now?"

"The last I heard," said Shorty, "he was making a machine that would allow a pianist to turn the pages of his music without using his hands. He whispered the idea one day to a pianist who is having trouble with arthritis, and the man has been working on the invention practically night and day ever since. With Freddie whispering ideas, of course."

Freddie mumbled something in his sleep and smiled.

Tom began to draw a picture of Freddie. He realized, listening to the woodspirits' conversation, that they were in that cheerful, expansive mood that often accompanies good food, good drink and good company. He shifted slightly where he sat—the ground was beginning to get a little hard—and continued to draw. He was trying to capture Eva's sweet face and Shorty's shy smile on paper, quickly, lest he blink his eyes and all of the woodspirits be gone.

Tom took advantage of a lull in the conversation to ask if Freddie was likely to sleep all night.

Hyke laughed. "Freddie loves to sleep even more than most woodspirits, and heaven knows we all enjoy it! He'll wake up eventually, I suppose . . . whenever he is ready."

Tom shook his head. "I'm not sure I could sleep like that, though sometimes I need it. . . . In fact, some nights I pace the floor, just trying to sleep like Freddie does but having no luck at all."

Hyke nodded. "But you are so anxious about the next day, or perhaps the one just passed, that you can't sleep, isn't that it? Well, woodspirits try to fall asleep each night in harmony with the rest of nature, so that sleep refreshes us and we wake up with

nothing but enthusiasm for the new day. If we do have problems falling asleep, then there is a woodspirit whose special job it is to go to those few woodspirits who are unable, for some reason, to make peace with themselves before they go to sleep, and he helps them resolve the problem as they lie there in their beds. They usually don't even know he's there — they just wake up the next morning knowing what they must do to be calm and happy again, and they go apologize to another woodspirit, or they finish that task that was troubling them the night before."

Eva added, "I'm told that humans have their own ways of seeking peace and calm . . . some read, some make themselves relax by breathing certain ways, or doing — what do you call them? — calisthenics. Unfortunately, many humans seem to turn to food, or drink, or herbs or potions of some kind to help them sleep. Those are the ones who would benefit from the ministrations of our special woodspirit of sleep, or who should learn a lesson from Freddie here!"

Freddie turned over, sighed peacefully and continued to slumber while the other woodspirits told Tom stories about their days as youngsters, about their history and about their dealings with humans. They also spoke of animals they had known, and beautiful seasons of the year. They reminisced about the area around Sink's Farm, long before Tom had even been born, and they spoke with love of woodspirits long since passed into final rest.

Tom continued to sketch as the woodspirits talked. He tried to put the creatures'

beautiful, wrinkled faces on paper, and the glow of the small fire they had made, and the way the tree branches seemed to stretch down towards the clearing in search of companionship, just as the woodspirits drew closer together as the night grew darker. Once Tom looked up and saw sparkling stars; another time he let himself drift very gently into sleep, then jerked himself back to awareness, afraid that his time with the woodspirits would come to an end before he was satisfied with his sketches of them. He scribbled; they talked.

Finally, a wind came up suddenly, as night winds sometimes do, and wove itself around the trees and chilled the woodspirits as they sat in companionable silence around the fire. The flames sputtered and died. Even Freddie finally awoke and rose to help fan the coals back into life, and soon the branches were crackling and snapping brightly and peace was restored.

Tom looked up from his sketchpad to find Hyke next to him. He turned his pad so that the woodspirit could see the drawings, and Hyke peered at them in the firelight, nodding appreciatively.

"You have captured us on paper, I see," Hyke told his friend. "You are quite talented, Tom."

"And you, Hyke, are quite well-loved," Tom answered, taking his sketchbook and putting it in his pack. "You must be very special to have friends like these around you on your birthday."

Hyke looked pleased. In fact, Tom could have sworn that the little woodspirit actually blushed, but in the rosy light cast by the fire it was hard to be sure.

"Please go on with your party, Hyke. I'm afraid that it's still a little chilly for an old human like myself, so I'll be going now," Tom told his friend. "Please—don't show me the way! I know my way home now. Stay with your friends and celebrate!"

Hyke nodded. "Thank you for coming, Tom, to my party! I know my friends enjoyed meeting you, too, or they would not have spoken so freely of old times in front of you . . . "

" . . . or made so many jokes about humans!" Tom laughed. "I can't blame them. We're a funny bunch, aren't we?"

Hyke nodded again and smiled, then lifted up his acorn cup. "I know that humans and woodspirits have several things in common, and one of those is that when we have special occasions to celebrate, we propose a toast. We've kept a Latin phrase from ancient days to describe our woodspirit toasts—'Esto perpetua.'"

Tom lifted up his own small cup. "Let's see, if I can remember some of the Latin I learned in high school, that would mean 'May you live forever,' or something similar."

"That's right," said Hyke. "Perhaps we

should add, 'May you live in peace and happiness with nature and your fellow creatures,' for in woodspirit terms that is the definition of a long and happy life."

Hyke added, "Let us also toast you, Tom, and the friendship between us and you," and he pointed to the other woodspirits, still talking (or, in Freddie's case, sleeping) in front of the fire.

"And let us toast Shady, who first brought us together," Tom responded. "You told me then that you came when you were needed, and I certainly needed you then."

Hyke smiled. "And I have benefited from our meetings, too," he told Tom. "Woodspirits like to receive as well as to give."

The two friends drank (or pretended to drink—neither one had any beverage left in his cup) and then Tom got to his feet very slowly and carefully, so as not to disturb the other woodspirits as they sat in front of the fire. He stretched, lifted up his pack and then squatted down and spoke one more time to Hyke.

"See you soon? As in Tuesday evening at seven?"

Hyke grinned and nodded. "Tuesday at seven it is. I'll help you continue your drawings, shall I?"

Tom nodded and quietly made his way out of the clearing. He stopped once and looked back, but the thick bushes and brambles covering the ground concealed the woodspirits from his sight. All he could make out was the faint orange glow of the fire, and the outline of a woodspirit's pointed hat. Then Tom turned and began his walk home.

He took a longer way back to the house than usual, partly because the night was so beautiful and clear, and partly because he wanted time to think about the woodspirits. Somehow it was easier to think about them in their own setting—nature—than in his house, where human thoughts and clutter sometimes made him forget everything else. As he walked along the path that led from Sink's Farm to the main road that passed his own home, Tom thought about Eva and Shorty and Skipper and Hyke and yes, even shy Franklin and the slumbering Freddie (who no doubt was just now waking up and preparing to go about his nightly duties).

What was it about the woodspirits that was so unique, so . . . well, woodspirited? Tom thought about their bright eyes, their wrinkled faces, their wide smiles, and he remembered how they moved: not too slowly, not too quickly, just the right speed with which to reach their destinations, be it Hyke's kitchen or the nearest plate of goodies. They didn't try to impress each other, or be the center of attention, or talk with one woodspirit to the exclusion of another. They didn't rush into the clearing talking of other places to go; they didn't complain about traffic, or papers to grade, or meetings to attend, or errands to run.

In fact, Tom realized he had never been around creatures so intent on enjoying themselves this minute, this place! Even though they talked about days gone by, it wasn't with regret or longing, as Tom sometimes heard himself talking about the past, but with thankfulness and happiness. The future seemed to be viewed in just that manner: how wonderful if indeed the next day, or month, or year did come, but if it didn't, well, one had had plenty of wonderful days already.

Tom found himself standing in front of Shady's grave, and wondered if in fact his heart had been leading him to just this spot, even as his head directed his steps home! He looked down at the earth and thought about Shady, and when he remembered his happy times with his dog he was happy himself.

"I'm not grieving anymore," Tom said to himself. "I miss Shady, but I'm not angry because she left me all by myself like I used to be, and I don't feel like I'll never smile again . . . "

He looked up at the stars, then spoke aloud to Shady. "I might even adopt another dog someday. What would you think of that, girl? She wouldn't be as wonderful as you, but I could talk to her about you and take care of her and give her a good home, and she could be friends with Hyke just like you were . . . "

He stopped. Suddenly the peacefulness of the moment grew even stronger, just as the air seemed suddenly warmer and more gentle, and the stars seemed to be trembling, on the verge of saying something to him. For the first time, Tom caught the scent of spring on the wind, and maybe—very faintly—the scent of the woodspirits' fire, way off in the distance. The statue of Hyke he had placed at the grave even looked as if it was smiling at him.

Tom didn't question anything he saw or heard or felt just then. Too many wonderful, inexplicable things had happened lately to him to make him question anything! He simply realized that he would never have been ready for this spring if it hadn't been for Hyke, and the things Hyke had taught him, and suddenly he was more eager than ever to continue his sculpting of the little woodspirit. Maybe he could even sculpt Skipper, and Franklin, and Eva, and Freddie, and Shorty, too.

But however many woodspirits he decided to capture in clay, and however many days it took him, he would enjoy them. He would stop, just as woodspirits did, to savor every mark he made in the clay, every minute he spent in his beloved studio, every walk he took past Shady's grave.

And that wasn't all. Maybe, just maybe, other human beings would like to know about Shorty and Hyke, Eva and Freddie. It was possible, wasn't it, that other human beings had seen woodspirits? And maybe there were many other people out there who were looking for something and weren't quite sure what it was, and Tom could tell them about Hyke and they would feel the way he did right now. Maybe, just maybe, it was possible to capture in clay the comfort and encouragement he'd been given by Hyke and the other woodspirits. His art had never let him down before, thought Tom. No reason why his art couldn't speak for him again, now that Hyke was in his life.

Tom looked down at Shady's resting place again and thought he saw the edges of a violet peeking through the damp leaves. Perfect for woodspirit juice, he thought, and smiled to himself. "Here I go again, talking about woodspirits as if they were an everyday occurrence," he chided himself, and then he realized that since meeting Hyke, woodspirits had indeed been part of his daily existence . . . or rather, the ideas of woodspirits had been. He kind of liked being able to talk and think like woodspirits talked and thought, and he realized that if any other creature understood, Shady did.

And now spring was finally coming. And Tom knew he was ready to go back into his studio and try to capture the spirit of his new friends in clay for other human beings to enjoy.

Smiling, Tom said farewell to Shady and walked on down the path to his studio, where a light was shining from the window to welcome him home.

THE END

# A BRIEF HISTORY OF WOODSPIRITS

Woodspirits through the ages have been confused with other creatures, and therefore called a variety of names, some of which they don't mind and others they do.

Greek mythology contains references to Napaeae, Auloniads, Hylaeorae and Alsaeids, who were described as being nymphs or spirits of the wood. Dryads—forest nymphs—are also mentioned in Greek tales and legends, while the Romans spoke of Oreads, who were nymphs of the mountains and hills. All of these references are actually to woodspirits.

Once woodspirits began to migrate out of Italy and Greece to central and then northern Europe, they accumulated another set of names. Teutonic mythology mentions a race of little people known as "forest spirits" who were "connected with the yearly renewal and decay of Nature." One of the customs in that region was to carry home young trees and green shoots during early spring, planting them outside one's home so that any human who came in contact with the new growth would also absorb the newly revitalized personality of the forest spirits. These spirits also were called "witugeist," which in Old High German means "wood spirit."

In northern Europe, Swedes called woodspirits "skogsfruar," or wood nymphs, while the Danes referred to them as "askefruer," or ash nymphs (ash being the predominant tree in Denmark). Both cultures described the creatures as having wrinkled faces and wearing garments of moss, a natural covering still used by modern woodspirits when they want to escape human detection in the forest.

Throughout the ages, woodspirits have often been confused with their more abundant and less publicity-shy cousins, the gnomes, whose name appears as γη-γενης ("earth dweller") in Greek, "gnomus" in Latin and "gnomen," or "earth dweller," in Middle English.

## Origins and Travels

From the sheer multitude of Greek and Latin words that are used casually and naturally by modern woodspirits, historians assume that the creatures were on earth at least that long ago, if not before. But most tales passed from one woodspirit to another concern later years, when woodspirits (with their cousins the gnomes) joined a huge expedition of little people from the Mediterranean. These creatures made their way out of those ancient lands into the part of the world now known as central and northern Europe.

A number of woodspirits did choose to remain in Germany, France, Holland and Austria (and their descendants are there still, so far as we know). But the majority traveled over the narrow waters separating central

ROUTE OF WOODSPIRITS TO THE NEW WORLD

Europe from the British Isles and made their homes there. This migration probably took place around A.D. 449, when the Romans lost their rulership over Britannia to the Anglo-Saxons and Jutes. As the Romans fled Britain, woodspirits were going to Britain, sometimes in tiny ships of their own making, more often as stowaways on fishing boats.

Once in Britain, woodspirits proceeded to become very much a part of Celtic legend, though those fanciful people often mistook woodspirits for fairies or elves or even gnomes (who did not actually go into Britain until fairly recent times, their personalities being more in tune historically with the Teutons

than with the more mercurial Anglo-Saxons).

Though Celtic legends do group woodspirits with wights, or elves, woodspirits were actually happier taking care of the woodlands and streams in Britain than they were building elvish barrows or helping to erect the huge stones that were used by Celts in worship ceremonies and that still attract tourists to Stonehenge and Keswick.

One project taken on by woodspirits involved the care of mazes, which were intricate patterns made from low bushes and trees such as the yew. Unlike the ancient Greeks, who built mazes primarily to house evil spirits

[57]

(such as the terrible Minotaur of Crete), the Romans saw maze-building as a sort of mental exercise. They did not enjoy the planting and cultivating of the bushes half as much as they enjoyed the designing of complex and intricate mazes, and so by the time the woodspirits came, most of the Roman mazes had fallen into neglect.

The woodspirits, however, loved trees and bushes and immediately set about clearing the mazes of weeds and restoring them to their original state. Some, in fact, still exist today, thanks to the ministrations of wood-spirits, and are still visited by tourists who marvel at the intricately planted boxwoods and yews. The most popular ones—"Troy Town" on the Scilly Isle of St. Agnes and a maze in Essex, on the Saffron Waldron Common—simply exist now for the viewing pleasure of humans, which makes woodspirits very happy.

Woodspirits lived happily in Great Britain for hundreds of years, despite being a fairly unknown group when compared with the more visible elves and fairies of the region. Still, when English ships began to set off across the wide ocean for the New World, the woodspirits' natural curiosity got the better of them and they, too, decided to seek their fortunes in this tantalizing unknown land.

Therefore, a group of woodspirits appealed to a group of humans with whom they had had unusually close contact—the Quakers—and were given permission to accompany the Quakers to the New World. This was in the year 1682. The leader of the settlers was William Penn, a peace-loving Quaker who with quiet zeal had recruited like-minded individuals to join him in establishing a settlement across the ocean from Europe, where they would be free of religious persecution.

Though references to the woodspirits do not appear in any of the ship's logs, tradition has it that the woodspirits were given fairly comfortable quarters in the "Welcome" by Penn himself, who no doubt felt that woodspirit knowledge and skills might come in handy in the New World.

While woodspirits are peace-loving crea-tures—and have deep respect for and under-standing of the religious beliefs of humans—they are also adventurous and unusually (for Little People) open to new beliefs and concepts. Therefore, they liked the idea of leaving behind the ancient (and often staid) ways of their ancestors and founding their own new eisteddfod, one in which they would be free to pursue woodspirit hobbies and relieved of the advice of their cousins the gnomes (who tended to frown upon many woodspirit customs as "frivolous").

Accordingly, when William Penn's band of settlers landed in the New World and set up a town they called Philadelphia ("brotherly love"), the woodspirits enthusiastically set up their own community in the mountains and woodlands surrounding the town.

Because the woodspirits ventured outside the confines of their homes more than the somewhat fearful settlers did, they were the first to meet the Indians who lived in the area. A fairly cordial relationship developed between these Native Americans and the small creatures who called themselves woodspirits—not so unusual when one remembers that Nature played such a part in the philosophy of both woodspirits and Indians.

These meetings with the Indians were recorded by Gloucester, a woodspirit from Gloucestershire in England. He was the historian of his day, and his records of the woodspirits' life in the New World are both lively and informative.

For example, not only did woodspirits feel some kinship with Indians, but they also discovered that the diversity of Penn's settlers was appealing to them. Penn's hardy bunch traced their families back to Germany, France, Sweden, Austria, Ireland and Scotland. As a result, the language spoken by the Philadelphians was English sprinkled liberally with foreign words and terms, a practice also adopted by the woodspirits. The colorful garments of other cultures, along with unusual customs and ceremonies, were also adopted by both settlers and woodspirits. And unlike some of the other New World groups, who became quite narrow and rigid in their beliefs, the settlers of "melting pot" Philadelphia accepted rather than rejected strangers.

The woodspirits of the New World found that they also admired many customs practiced by the Indians of the area, most of whom were of the Algonquin tribe. Both the woodspirits and the Algonquins shared a belief that everything in Nature—humans, Little People, plants, stones—is inhabited by a mysterious power, which spreads out and influences other beings. The Indians had various names for this power—"Orenda," "Manitou"—but the belief in an intelligent and powerful force behind nature was the same.

In those early days of America, both Indians and settlers benefited from the activities of woodspirits. Seeds brought to the New World by the settlers flourished in their new surroundings, thanks to woodspirits; cows produced more milk, thanks to the encouragement of woodspirits; wells were dug by dowsers, humans who were supposed to have miraculous abilities to detect fresh water supplies below ground but were in fact simply led to those spots by their friends, the woodspirits. And woodspirits showed Indians in the area how to apply some of the white man's inventions, such as gardening tools and wax candles, to their own way of life.

Still, the woodspirits often found themselves caught between two cultures, the Indian's and the white man and woman's. As more settlers arrived in the New World, the Indians began to feel threatened, and skirmishes broke out. While settlers were killed by angry Indians, angry settlers killed or tortured Indians. The woodspirits watched the battles helplessly, hoping that eventually the Indians and settlers would be able to coexist in peace and harmony.

As the settlers in Philadelphia began to lay their town streets, build their houses and establish commerce, woodspirits found their services less in demand than before. Besides, they were curious to see what lay south of the Adirondacks. And so, in small bands (leaving some of their number behind to continue the woodspirit line), woodspirits began to make their way south, some settling in the areas now known as the Chesapeake Valley, some stopping for a time in what is now Virginia, and many more traveling on over the Appalachian mountains into the Carolinas.

In these regions, many woodspirits found the temperate climates they loved (parts of Europe having been too cold and parts of Greece and Italy having been too hot for their liking). They found, too, the forests, which have long held special appeal for woodspirits, offering as they do shelter, tools for survival and esthetic beauty. Here, too, were vast amounts of herbs and flowers, which woodspirits used as healing potions, food and inspiration (and still do today).

## Woodspirits in Literature

One can only wonder if perhaps Shakespeare's Puck, Jonathan Swift's Lilliputians or the leprechauns of Irish folklore were, in fact, woodspirits. In any event, no mention of woodspirits is made in English literature until the Romantic Era, when woodspirits were grouped with elves, dwarves, fairies and their cousins, the gnomes, in some obscure tales and legends. Many English-language encyclopedias do not mention woodspirits at all and gnomes very infrequently. Lang's *The Book of Fairies*, in fact, calls both gnomes and woodspirits creatures that have resulted merely from "the romantic elaborations . . . of writers."

However, woodspirits are mentioned in an indirect way by writers of other countries. The Brothers Grimm wrote a story in which a gnome—or perhaps a woodspirit—helped a young boy rescue the daughter of a king from an underground monster. A book on Rhine River legends speaks of a group of gnomes—or woodspirits— helping a young boy build a road, a somewhat odd task for which his reward was the opportunity to marry a princess.

## Woodspirits Compared to Other "Little People"

Woodspirits may often be confused with other Little People—elves, dwarves, gnomes—but, in fact, they have distinct personalities, habits and preferences.

The Dutch book about gnomes, which appeared in the 1970s and was an immediate best-seller, describes gnomes as dressing almost exclusively in blue smocks and red caps. Their cousins the woodspirits, however, have always been quite fond of different garments, especially those that allow them to blend, when they wish, into their natural habitat. For example, a yellow coat may be worn by a woodspirit who lives among daffodils in the spring, while a hat made of a leaf lets him mingle with the autumn leaves covering the ground.

Other differences between woodspirits and gnomes concern habits as much as anything. While most female gnomes stay at home with their offspring, female woodspirits are quite likely to have trades outside the home, with care of any offspring being handled by the "extended family" that woodspirits take pains to sustain. Also, woodspirits are much more eclectic in their choice of dwellings: where a gnome prefers an underground home, woodspirits will make wonderful dwellings out of trees, haylofts, deserted cars and other above-ground structures.

Woodspirits are even less like elves, with whom they are often confused, than like gnomes, who at least are cousins. Elves distrust iron and steel, industrialization and religion; cannot count; and rarely know the names of the days of the week or any other way in which humans measure time. While elves occasionally help humans by easing the pain of childbirth or lessening the fury of thunderstorms, most of the time they are willful and spiteful creatures, almost as moody as Mother Nature herself.

On the other hand, woodspirits are known for their sunny dispositions, their stable personalities, their curiosity, their skill in anything requiring precise handiwork and their love of travel. They are also generally admiring of human inventions and machines, though not always pleased with the ways in which those machines are used.

## Talents of the Woodspirits

Woodspirits have always been good, skillful gardeners. One of their particular abilities, stemming from their long and patient observation of nature, is determining which plants do well planted together and which ones can harm each other. For example, lilies of the valley and narcissi grown together soon wither, woodspirits say, while rue and basil kill each other when they are planted together in the same herb garden. Likewise, clover and buttercups, and mint and parsley, are bad mixes.

These tidbits of herbal lore, when demonstrated to a certain Quaker fancier and doctor, John Bartram, actually inspired him to plant the first botanical garden in the New World, in 1730. It originally occupied five acres on the Schuylkill River in what is now known as the state of New York; today, the same garden is watched over by woodspirits specially trained by their ancestors in botanical gardening and herbal lore.

Woodspirits all around the world have continued to inspire humans to set up communities in which gardening is the focus. The most famous such community, Findhorn, was founded in Great Britain in the 1960s. Members of that community have published well-known and respected books saying that mystical beings called "devas" (or woodspirits) are responsible for the cooperation between man and nature that has caused Findhorn to prosper. Similar communities in America have also prospered, thanks to woodspirits.

[63]

Thanks to their keen sense of smell, woodspirits have for centuries been excellent perfumiers, a talent they developed in Greece and Rome and eventually brought with them to Europe and, later, America. Woodspirits helped ancient Romans honor the reappearance of fertile spring with annual displays of flowers and plants. When woodspirits demonstrated these displays to the Britons, those humans happily accepted the practice and took part in it each year (except in areas where Puritans were in power; these somewhat rigid folk had banned the practice beginning in 1644).

The New World proved much more widely accepting of ceremonies using flowers, and woodspirits once again decorated homes, wells, churches, forest clearings, even urns with the most beautiful, fragrant plants of spring.

But woodspirits are not just gardeners and decorators: They are also of a mechanical mind. Their energy and curiosity were put to good use in the New World, where humans were struggling to find ways by which life could be made easier and more productive. While no humans of that time have written of their ties with woodspirits, there is no doubt that woodspirits were able to communicate ideas to them, either through whispered suggestions, words written on scraps of paper left lying about, or demonstrations.

Woodspirits themselves often took human inventions one step further, using the improved invention in their own small dwellings. For example, woodspirit homes in the New World of the 1700s had dripless wax candles, an improvement on the candles first made by humans sometime between 1,200 and 1,500 years after the birth of Christ. Woodspirits use special microscopes to detect and subsequently correct defective growth in plants; special pressure cookers to remove shells from acorns for use as bowls and cups; and special thermometers to test the amount of winter sunlight absorbed by garden soil so that spring buds can be predicted.

In fact, woodspirits have performed a number of tasks or perfected a number of inventions for humans that humans can only guess at, for almost always woodspirits prefer that humans see the results of what they do rather than set eyes on their small selves. For that reason, many soldiers wounded during battle have put their healing by a small creature down to delirium, when they would speak of such things at all. Musicians unable to get a certain bar of music quite right have told no one that the humming of some small creature helped them with their compositions. Artists have thrown down their brushes in disgust, only to come back the next day to find their sketches improved, as if by magic. And writers have rarely admitted to themselves that a line of poetry scribbled in tiny handwriting on a scrap of paper has made all the difference to the cadence of their sonnets—for who would believe them if they said the handwriting was that of a woodspirit?

# THE MODERN-DAY WOODSPIRIT

### Size

Most woodspirits stand no taller than the measurement of a man's arm from elbow to wrist. The few who grow taller are said by their peers to be descended from a band of unusually hardy, vigorous and healthy woodspirits who traveled to the New World with the Vikings, rather than several hundred years later with the less robust William Penn.

### Shape

Pleasantly rounded in most cases, thanks to a love of eating. Woodspirits do not, however, dwell overly much upon how much they weigh or how round about the middle they are. They have perfected the art of eating well and in a leisurely fashion, yet never consume so much at one sitting that they are unable to move about in characteristically agile and speedy fashion.

### Complexion

Fair and wrinkled, though again woodspirits do not think too much about how they look. They prefer the rosy glow imparted by sunshine and breezes rather than by any chemical concoctions, and indeed look upon "character lines" in a woodspirit's face with admiration.

### Average Lifespan

Most woodspirits live between 350 and 400 years. This means that there are actually alive today some (very, very old) woodspirits who came into being during the time of the Renaissance, in Europe.

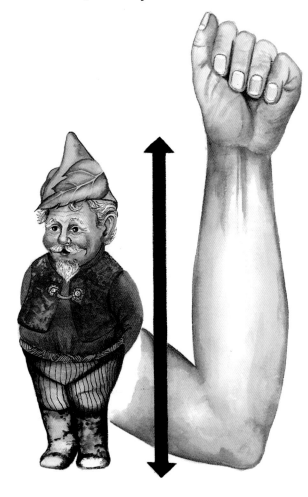

## Important Milestones in a Woodspirit's Life

The first twenty-five years are usually spent with one's parents or other designated family caretakers.

At the age of twenty-five, a young woodspirit then joins other woodspirit offspring in a community school, where he will customarily spend the next hundred years (seventy-five if he is quite, quite bright).

The teachers in the school are always the older woodspirits in the community, or eisteddfod. These venerable woodspirits are called the "old ones," not out of derision or pity but out of respect and love, and it is their job to instruct the youngest in the community in woodspirit ways. Other subjects are studied, too, along with woodspirit legends and behavior; these subjects include medicine, botany, psychology and mathematics.

In return for teaching, the old ones are given the finest dwellings in the eisteddfod by grateful friends and relatives, and are supplied daily with the tastiest food and drink. Having an old one to take care of, in fact, is considered a special mark of favor and blessing among woodspirits.

Reading, writing and memorization are employed in school, but equally important are the voyages made by the old ones and their pupils into the world. It is not unusual for an old one to lead a string of five, six or more young woodspirits along a path or over a hill, teaching the basic tenets of nature as they follow, for harmony with nature is of paramount importance to the survival and, in fact, prospering of woodspirit culture.

Coming of age for a woodspirit is not an overnight occurrence. He must first finish his formal education and leave the school in which he has labored for so long. Next, he enters a period of years—usually five—in which he is encouraged to travel to other eisteddfods for the purpose of talking with other woodspirits and learning from them. This is also the time in which the woodspirit works on an article of practical use as well as beauty for his future, adult home. Usually this article is a piece of furniture, such as a bed or table, and it is designed and made entirely by the woodspirit. Once a group of elder woodspirits has seen and approved the article, it is placed in the young woodspirit's new adult home and must be passed on by him to succeeding generations (or, if he has no offspring, to the eisteddfod itself). The article thus represents the young woodspirit

and the older, wiser woodspirit he is to be.

Once the article is approved by the community elders, a woodspirit next declares his trade. He has come to this decision based upon years of experimentation and observation, and chooses penmanship or the foretelling of weather or carpentry, knowing that his particular gifts and talents lie in this area. By having declared a trade, he becomes a vital, functioning part of the community.

Of course, many woodspirits—like humans—have more than one interest, and they are perfectly free to pursue those interests, provided that their primary trade has been practiced for the general good of his community.

## Family Life

Marriage is not even considered until a woodspirit is 125, and even then is only for woodspirits who have given long and careful thought to the responsibilities and privileges of marriage. Marriage for woodspirits is based not on any desire to increase one's status in the community or to acquire wealth, but from a desire to form a solid partnership with another woodspirit. About half of the woodspirits of marriageable age in an eisteddfod usually marry; the other half choose to affiliate themselves with an extended family in another form of partnership. Neither kind of partnership, community or marital, is considered better than the other.

Those woodspirit couples who desire offspring (about half of those in an average eisteddfod) are first counseled by their elders concerning their goal. Woodspirits love

youngsters, and yet they are the first to admit that rearing offspring is hard work, and not suitable for everyone!

Conception usually takes place in the spring. Gestation lasts until the following spring, at which time a number of woodspirit offspring will be born, gladdening the hearts of those in their eisteddfod. Once an off-spring is born, care may be undertaken solely by the parents, or in partnership with other woodspirits in the community.

## Clothing

Woodspirits love bright, colorful garments for ceremonies and practical, durable garments for trades. Just as human tailors were highly regarded centuries ago in Poland and other European countries, so woodspirit tailors are figures of respect, too. There is usually only one tailor for each five or six eisteddfods, and thus the tailor customarily travels between the communities, living for days with one before he moves on to another. During his

Woodspirits also love nothing better than swapping favorite garments with each other. It is not unusual for one woodspirit to give another a garment that that friend has particularly admired. Since all woodspirit garments are one of a kind, swapping clothes is a way of doubling the enjoyment.

## Dwellings

Woodspirits prefer above-ground homes to underground dwellings like those inhabited by their cousins the gnomes. Sunlight is important to woodspirits, and so dwellings have many windows, skylights and doors to allow light and air inside during fair weather. However, shutters and louvered doors of wood or heavy material (scraps of leather or cloth) allow the woodspirits' homes to be secure in times of cold or wet weather.

In ancient times, woodspirits used asbestos —thin fibers pulled from the stone serpentine—as wicks, a form of light still utilized by Eskimos in the cold, frozen North. Today, however, woodspirits' homes are lit by devices that use radio waves bouncing off of crystals, a technique only recently discovered by humans.

Walls are hung with art, some of it quite old, since woodspirits carefully pass paintings from one generation to the next.

Also providing decoration as well as function are great chests called cassoni (a word coined by southern Europeans in the 1500s, who adopted the woodspirit custom). The cassoni are decorated with intricate designs,

visits he will measure entire families of wood-spirits and sew special garments for them. These may include hats and coats that allow a woodspirit to blend in with his surroundings as well as the brightest, most colorful outfits for wear during special ceremonies.

Woodspirits believe that the color and texture of a garment increases the enjoyment of life of the one wearing the garment. For that reason, every special day on the woodspirit calendar (and there are many) is also a day when special clothing is worn.

usually employing flowers and birds, and are customarily used by woodspirits to store their garments and other important articles.

Beds are usually like those used by humans in Elizabethan times: large, curtained against drafts and having "truckle" (trundle) beds underneath to accommodate offspring or visitors. Chairs tend to be plumply upholstered; kitchens are hung with an amazing variety of appliances and gadgets, since woodspirits love anything that has to do with cooking, especially eating! The centerpiece of the home is likely, in fact, to be the kitchen, which usually holds a huge table for meals. Also central to most woodspirit homes is the study, where young woodspirits do their lessons and older woodspirits read and practice the fine art of conversation.

Plumbing in a woodspirit home is quite sophisticated. Woodspirits especially enjoy long, lazy soaks in tubs that are made by a specially trained tradesman to suit their particular sizes and shapes.

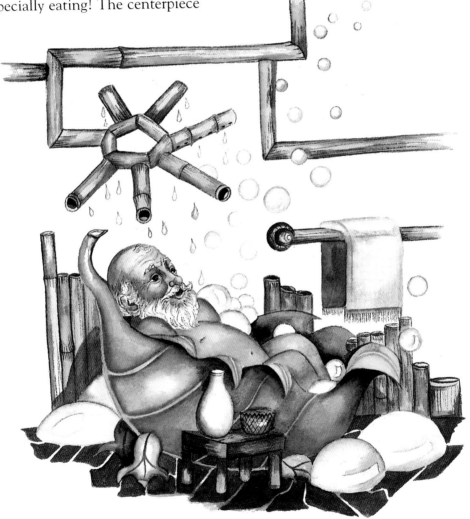

## Dining Habits

Meals for woodspirits tend to be happy, informal and frequent events. While there is no rigid schedule of mealtimes (woodspirits like to eat as the spirit moves them), still, there are certain times of day when a good number of woodspirits are likely to gather for eating, drinking and fellowship.

Woodspirits, in fact, have over the centuries perfected the art of dining, an art that was adopted by the Dutch in the sixteenth and seventeenth centuries. Usually at sunup, there is a light breakfast of bread, butter, milk and cheese. At noon, woodspirits may eat a light soup, fish of some kind, fruit and meat cooked in lemon juice. At sundown,

there is a supper consisting of tea, nuts, fruit, bread and cheese. And finally, for those woodspirits who truly enjoy eating or those whose trades keep them up late at night, there is a midnight snack of fruit and cheese.

In most eisteddfods, meals are not simply for the purpose of eating and drinking; they are also occasions for conversation. A large family may make attendance by all at mealtimes mandatory; smaller families may choose to take turns fixing meals for each other. Visitors, old ones and those with no families are always welcome at woodspirit meals. In fact, woodspirits originated the European custom of leaving bread, cheese and cordials on the table for any visitors passing by, and of leaving doors unlocked so that all woodspirits feel welcome.

## Trades

The trades practiced by woodspirits are just as diverse and customized as the clothes worn by woodspirits.

Some woodspirits will paint extra spots on fawns so that they will better blend in with their setting and so be protected from predators.

Others push baby birds back into nests when they teeter too far out on a tree limb; help the runt of a litter of pigs or puppies get his share of his mother's milk, which makes him grow stronger; and encourage hens (who tend to become easily bored) to sit upon their eggs until the eggs hatch by telling them stories and singing soothing songs to them.

Some woodspirits have a particular affinity for common household spiders. This affinity means that spiders will often weave especially strong ladders for tree-living woodspirits to use, while woodspirits will lead certain insects into a web, thereby helping perpetuate the ecological balance of nature. Often, so-called "writing spiders" will be used by older woodspirits to help teach offspring how to read and write (particularly attentive and open-minded human children will often be invited to view the work of writing spiders, too).

Some woodspirits twist the stems of apples so they will fall off the tree branches at the appropriate time, ready to be picked and enjoyed by humans, woodspirits and other creatures. The more daring and adventurous woodspirits will even straddle the apples and ride them down to the ground, using their pointed hats as parachutes so they don't land too heavily or bruise the fruit too easily.

There are woodspirits who keep birds and insects from colliding in midair; woodspirits who keep clocks in good running order; woodspirits who rock cradles ever so gently so that crying human babies will fall peacefully asleep; woodspirits who sharpen pencils for writers, oil parts for mechanics and generally make themselves useful in all kinds of handiwork.

Then there are woodspirits who specialize in delicate crafts outdoors, such as pushing stubborn new shoots through bulbs so they will bloom in the spring. Other woodspirits work with beavers to ensure that no stream is dammed up completely, or encourage squirrels to leave enough seed in the feeders to ensure that birds have food during cold weather.

# A FINAL NOTE ABOUT WOODSPIRITS

Woodspirits love to visit relatives, and though the number of woodspirits has dwindled —thanks to the increasing encroachment upon the land of humans and their machines —woodspirits still enjoy journeying across the land to see aunts, uncles, cousins, brothers, sisters and others.

Of late, on these journeys woodspirits have been shocked to see that the New World they entered centuries ago has changed. In fact, woodspirits are beginning to believe that as a group of living beings, they are in somewhat the same danger of being crowded out or made unwelcome as the eagles (though steps have been taken to preserve the lives of those noble birds). Many woodspirits are saddened to see that fertile farmlands are gone, quiet towns have become traffic-clogged, and cities and rushing streams are choked with debris and pollution.

As a result, many woodspirits have come back to their homes in the last decades determined to seek out understanding humans and tell them what they, the woodspirits, have seen. And so, while the actual numbers of woodspirits have lessened since that shipload of Penn's woodspirits set foot in the New World and multiplied, the numbers of woodspirits seeking human contact have increased.

That is one reason, perhaps, why Tom Clark's stories and drawings and statues of woodspirits have been so successful of late: woodspirits believe their customs and beliefs are being taken up by humans who are of the thoughtfulness and kindliness and energy to perpetuate these beliefs. And many humans have, indeed, shown themselves to be open to woodspirits and very grateful for the knowledge of Nature passed on to them by woodspirits.

Library of Congress Cataloguing-in-Publication Data
Grissett, Ellen F., 1955–
Woodspirits.
I. Clark, Tom (Thomas Fetzer)    II. Brown, Pamela
Rattray, 1945–   .   III. Title.
PS3557.R5365W6  1987   813'.54   86-28357